Cow Counting
1 ☆ 2 ☆ 3

by Claire Suminski
Illustrated by Susan Swedlund and Friends

We are very thankful for the following
contributing artists:
Pat Mennenger
Marilyn Miller
Ros Webb and
Lead Illustrator Susan Swedlund

First Edition
ISBN 978-1-7333559-9-5
Library of Congress Control Number: 2021930740

Published by Red Press Co.
Redpressco.com

Can you imagine trying to count all of the hairs on your head? Each one of us is so important to God. He has numbered the very hairs on our heads.
He loves each one of us so much!

But the very hairs of your head are all numbered.

Matthew 10:30

Nestled near the foot of Cowee Mountain in Western North Carolina is a little family homestead, called Cowee Mountain Valley Farm.

Chickens, angora goats, pigs, and steers are watched over by the farm's faithful Great Pyrenees, Cowee Sam.

Come count to twelve and back with us on the farm. Eggs and fresh baked farm cookies are counted by 12, which makes a dozen. When measuring, there are 12 inches in a foot. And there are 12 months in a year. Let's count!

1 Steer

2 Dogs

3 Cats

4 Eagles

Baby Owls 5

6 Pigs

7 Chickens

Goats

8

Bunnies

9

10 Opossums

11 Fireflies

12 Bees

So instead of counting...

1☆2☆3☆4☆5

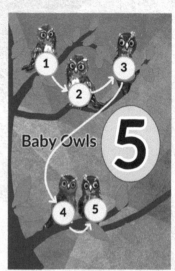

Count like this...

5☆4☆3☆2☆1

Are you ready to count backwards?

12, 11, 10, 9
8, 7, 6 ,5, 4
3, 2, 1

Let's give it a try!
We will help you
with the first four.

12 Bees buzzing

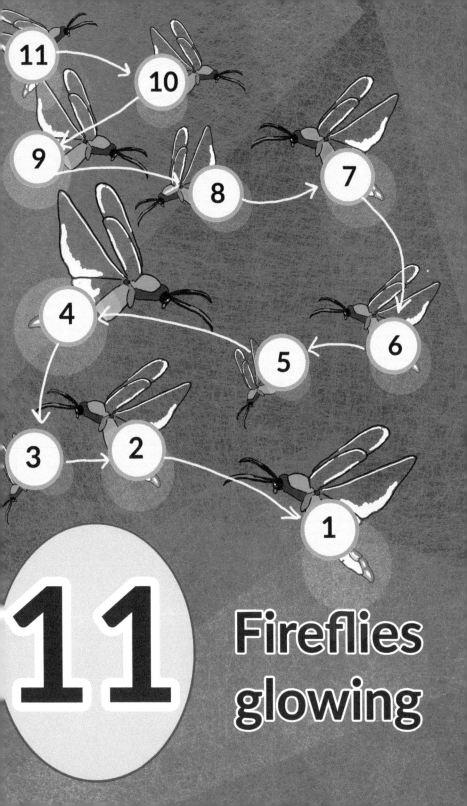

11

Fireflies glowing

10 Opossums climbing

Bunnies
hopping

9

Goats
munching

8

7

Chickens
cluck,cluck
clucking

6 Pigs oink, oinking

Baby Owls
hoo, hoo
hooting

5

4 Eagles nesting

3 Cats meowing

2 Dogs barking

1

Steer mooing

"Counting is an important building block for learning."

Grammy Ruby

Learning number recognition and practicing counting are ways to help your preschooler build number sense.

These building blocks are crucial to later success with math skills and sequencing. Counting is a part of daily life. Find ways to practice counting and make it fun for your child. Here are some ideas:

1. **Count fingers and toes**

2. **Count cans in the pantry**

3. **Line up stuffed animals in a row and count them**

4. **Collect eggs fron the hen house in the morning. How many are there?**

5. Find items in nature: rocks, leaves, flowers and practice counting them

6. Color in outlines of numbers

7. Make up your own number flash cards

8. Counting rhymes or songs

9. Put shaving cream on a cookie sheet and practice drawing the shapes of numbers with the pointer finger, then smooth out the shaving cream with a rubber spatula and draw more numbers

Counting helps brain development and it is fun!

Try this fun rhyming song from the early 1800's:

1,2 Buckle my shoe
3,4 Shut the door
5,6 Pick up sticks
7,8 Lay them straight
9, 10 A big fat hen